# CHEMISTRY

## BULLET GUIDE

### Helen Harden

Hodder Education, 338 Euston Road, London NW1 3BH

*Hodder Education is an Hachette UK company*

First published in UK 2011 by Hodder Education

This edition published 2011

Copyright © 2011 Helen Harden

The moral rights of the author have been asserted

Database right Hodder Education (makers)

Artworks (internal and cover): Peter Lubach

Cover concept design: Two Associates

*British Library Cataloguing in Publication Data:* a catalogue record for this title is available from the British Library.

10 9 8 7 6 5 4 3 2 1

The publisher has used its best endeavours to ensure that any website addresses referred to in this book are correct and active at the time of going to press. However, the publisher and the author have no responsibility for the websites and can make no guarantee that a site will remain live or that the content will remain relevant, decent or appropriate.

The publisher has made every effort to mark as such all words which it believes to be trademarks. The publisher should also like to make it clear that the presence of a word in the book, whether marked or unmarked, in no way affects its legal status as a trademark.

Every reasonable effort has been made by the publisher to trace the copyright holders of material in this book. Any errors or omissions should be notified in writing to the publisher, who will endeavour to rectify the situation for any reprints and future editions.

Hachette UK's policy is to use papers that are natural, renewable and recyclable products and made from wood grown in sustainable forests. The logging and manufacturing processes are expected to conform to the environmental regulations of the country of origin.

www.hoddereducation.co.uk

Typeset by Stephen Rowling/Springworks

Printed in Spain

*With thanks for the continued love and support of my husband, Paul,
my two wonderful boys, Francis and Timothy, and my parents*

## About the author

Helen Harden specializes in the writing and development of science curriculum materials, in particular for the Association for Science Education and the Twenty First Century Science project, as well as other publishers. In addition, she periodically runs workshops at the Science Learning Centre London to help staff in school science departments develop their own teaching aids.

She was previously a teacher at a comprehensive school in Hertfordshire where she became head of its chemistry department. Helen lives in Hertfordshire with her husband and two boys. In what little time remains after family and work, she sings with the Hertfordshire Chorus and Voix de Vivre chamber choir.

# Contents

# Introduction

Chemistry is the study of materials – the air we breathe, the wood, metal or plastic of the chair we sit on, or the diamond of a ring we wear. Chemistry allows the **synthesis of complex molecules** to develop life-saving medicines. Chemistry can make a firework explode spectacularly or slow the decay of an ancient metal object. In short, chemistry is the science of everything.

Those that study chemistry learn to see **the world of the minute**, a world that goes unseen by most others. They can use their understanding of **invisible atoms and molecules** to explain visible observations such as the changing colours in reactions. The chemicals inside the bottles and jars in a laboratory may be measured, weighed and poured to **synthesize molecules never before made by humankind**, while the latest technology can be used to **analyse and identify previously unknown compounds**.

vi

## In short, chemistry is the science of everything

For readers in the UK, this book **bridges the gap** between GCSE and AS-level. It is suitable for anyone who has studied GCSE or O-level who wishes to **refresh and extend their understanding**. It would also make a **valuable study aid** for any students starting an AS-level course.

The contents of this book are by no means all you need to know, but they are certainly things you should not go into an exam *not* knowing.

vii

### 'Better Things for Better Living Through Chemistry.'
Advertising slogan for DuPont, 1935

# 1 Atoms

# Understanding atoms

For centuries, philosophers and scientists alike have puzzled over the question:

## 'What are substances made from?'

The theory that all substances are made of atoms, which are the smallest possible indivisible particles, remained undisputed until the twentieth century when even smaller subatomic particles were discovered.

*How do you age the body of a man that has been preserved in a bog for hundreds or perhaps thousands of years?*

Scientists in the twentieth century discovered that atoms consist of a **nucleus** surrounded by **electrons**.

It is the **electron arrangement** that determines chemical reactivity and is therefore of greatest interest to chemists.

This chapter outlines how to work out the number of protons and neutrons of different elements and possible models to describe the arrangement of electrons.

It details the workings of a **mass spectrometer** and explains how to use isotopes to calculate the relative atomic mass of an element.

A case study summarizes how isotopes help to date archaeological remains.

# Subatomic particles

Atoms are made up of three kinds of **subatomic particles**:

✳ **protons** and **neutrons** are found in the central **nucleus** of the atom
✳ **electrons** are located around the outside of the atom.

## Table 1.1: The three kinds of subatomic particles

|          | Relative mass | Relative charge |
|----------|---------------|-----------------|
| Proton   | 1             | +1              |
| Neutron  | 1             | 0               |
| Electron | 1/1840        | −1              |

✳ The **atomic number** and **mass number** can be used to work out the number of each type of subatomic particle.
✳ The periodic table gives the atomic number (smaller) and mass number (larger) for each element.

4

* The **atomic number = number of protons**.
* As the **number of electrons = number of protons**, so the atomic number also gives the number of electrons.
* The **mass number = number of protons + number of neutrons**.
* The **number of neutrons = mass number − atomic number**.

**Worked example**

How many protons, neutrons and electrons are there in a helium atom?

The mass number of helium is 4 and its atomic number is 2.
Number of protons = atomic number = 2
Number of electrons = number of protons = 2
Number of neutrons = mass number − atomic number = 2

HC.

# Electron arrangements

Scientists choose the most suitable **model** of electron arrangement to work with. A model is a way of thinking about a phenomenon that helps to answer a problem:

* 'Dot and cross' diagrams, which represent the location of electrons in basic, circular **electron shells**, are good for working out the bonding between atoms.
* The use of **s**, **p** and **d subshells** is helpful in explaining trends in the periodic table.
* Treating electrons as forming a **continuous electron cloud** is useful in describing the properties of materials. **Electron density** is used to describe the spread of electrons within a molecule.

6

**TOP TIP**
Remember that the first electron shell holds two electrons; the second and third can each hold eight electrons.

To create a basic 'dot and cross' electron diagram for magnesium (which holds 12 electrons)…

1 draw three concentric circles
2 label the centre Mg (magnesium)
3 start adding electrons by drawing dots or crosses
4 always start in the inner circle
5 fill this first shell by adding two electrons
6 add eight more electrons to fill the second shell
7 add the remaining two electrons to the third shell.

### Note!

The outermost electron shell is most stable if full. Magnesium can lose the two outer electrons to become a charged ion ($Mg^{2+}$), leaving a full outer shell.

# Isotopes

While the number of protons defines the element, the number of neutrons can vary.

Atoms with the same number of protons but differing number of neutrons are called **isotopes**. It is the number of electrons that determines the chemical reactivity of an element, so different isotopes will react identically. Elements can exist in a mixture of isotopes but usually one isotope predominates.

8

### Carbon isotopes

Carbon exists in three isotopes. The most abundant isotope is known as carbon-12, where 12 signifies the mass number. Only 1.11% of carbon is found as carbon-13, with trace amounts of carbon-14 also present.

## CASE STUDY: Bog Man

In the 1950s the amazingly preserved remains of a man were found in a bog in Ireland. At the time other circumstantial evidence was used to estimate his age as about 2,000 years old. Later the newly discovered method of radiocarbon dating gave a date of around 400 BC.

Radiocarbon dating works because…

* tiny amounts of carbon-14 are absorbed by all living things
* carbon-14 is a radioactive isotope that gradually decays
* once a living things dies, no new carbon-14 is absorbed
* the existing carbon-14 gradually decays away
* scientists can measure how much carbon-14 is left
* they know its rate of decay and can therefore work out the age of the sample.

# Mass spectrometry

A mass spectrometer may be used to help identify an unknown molecule.

## Table 1.2: How a mass spectrometer works

| Step | Part of mass spectrometer | Function |
|------|---------------------------|----------|
| 1 Vaporization | – | Sample already vaporized |
| 2 Ionization | Electron gun | Bombards the gaseous sample, knocking out electrons and creating **positive ions** |
| 3 Acceleration | Charged plates | **Accelerates** the ions |
| 4 Deflection | Controllable magnetic field | **Deflects** the ions (the amount of deflection depends on the mass of the ions) |
| 5 Detection | Detector | **Detects** and records the relative abundance of each ion's mass |
| 6 Results | Read-out | Records the result and is known as a **mass spectrum** |

## Relative atomic mass

Periodic tables often give a mass number that is not a whole number – this is the **relative atomic mass (RAM)** of the element.

### Worked example

Calculate the relative atomic mass of chlorine.

The RAM takes into account the existence of different isotopes. The relative atomic mass may be calculated by working out the average mass of the chlorine atoms.

The mass spectrum of chlorine reveals that about 75% of chlorine atoms have a mass of 35 and 25% have a mass of 37.

You can work out its average mass by considering 100 chlorine atoms:

$$\frac{(75 \times 35 + 25 \times 37)}{100} = 35.5$$

# 2 Bonding

## The importance of bonding

The properties of the chemicals around us are essential not only to twenty-first century living but also to life itself.

Yet, invisible to us, it is the type of **bonding** within these chemicals and their structure that determine these individual properties.

*Why does an iceberg float?*
*Why don't Earth's oceans boil?*

### The properties of the chemicals around us are essential... to life itself

● Bonding is one of the most important fields of chemistry.

Salt dissolves, copper conducts electricity and oxygen is a gas. These properties arise from both the type of bonding – **ionic**, **metallic** or **covalent** – and the overall structure – **giant** or **simple molecular**.

In this chapter we will cover:

* the three main types of bonding
* the resulting structures *and*
* the forces that occur between molecules.

This chapter also illustrates the importance of **intermolecular forces** to life itself by considering the structure of DNA and proteins.

# Types of bonding

## Ionic bonding

* Ionic bonding occurs between **metals** and **non-metals**.
* The metal atoms **lose** their outer electrons and become **positively charged** ions.
* The electrons are **transferred** to the non-metal atoms.
* The non-metal atoms become **negatively charged** ions.
* Both positive and negative ions have **full outer shells** of electrons.

16

> ### KEY FACT
> Sodium chloride has a **giant ionic lattice structure**.
> The positive sodium ions are attracted to the negative
> chloride ions by **electrostatic** forces.

## Metallic bonding

Metals have distinctive properties such as being strong, shiny, malleable, ductile and conductive to heat and electricity.

In a metallic structure, the metal atoms still lose their outer electrons but there are no non-metal atoms to receive them. Instead, the outer electrons are **delocalized** between the positive metal ions. The ability of this electron cloud to move accounts for the **conductivity** of metals of both heat and electricity.

**Metals have distinctive properties such as being strong, shiny, malleable, ductile and conductive to heat and electricity**

# Types of bonding/structure

## Covalent bonding

* Covalent bonding occurs between **non-metals**.
* Electrons are **shared**.
* Each atom gains a stable **full outer shell**.
* If just one atom provides the **pair** of electrons, a **co-ordinate (or dative) bond** is formed.
* Where two electron pairs are shared between atoms, a **double bond** is formed.

## Structure and bonding

The properties of a substance depend not only on the type of bonding but also on the overall **structure**.

**Table 2.1: Properties related to bonding and structure**

| Bonding | Structure | Properties | |
|---------|-----------|------------|------------|
| | | Melting point | Electrical conductivity |
| Ionic | Giant | High | Liquid and aqueous solution |
| Metallic | Giant | High | Solid and liquid |
| Covalent | Giant | High | Do not conduct |
| | Simple molecular | Low | Do not conduct |

An exception is graphite – its structure is made up of layers of carbon atoms. One electron from each carbon atom is **delocalized** between the layers. This enables graphite to conduct electricity.

# Bond polarity

* In a covalent molecule where both atoms are the same, the electron density will be **symmetrical**.
* If one atom is more **electronegative** than the other, there will be greater **electron density** near the more electronegative atom as it pulls a greater share of the electron cloud towards itself.
* The covalent bond will be **polar** and a **dipole** will be created.
* The greater the electronegativity of the atom, the more polar the bond.

Electronegativity increases as you go across the periodic table, as shown here in period 2:

lithium → beryllium → boron → carbon → nitrogen → oxygen → fluorine

The slight negative charge ($\delta^-$) on the end of a polar molecule will be attracted to the slightly positive charge ($\delta^+$) on another polar molecule. This results in a force of attraction **between** the molecules.

* **Intramolecular** bonds **within** a molecule are very strong.
* Weaker **intermolecular** forces of attraction exist **between** molecules.

Figure 2.1: The three types of intermolecular forces

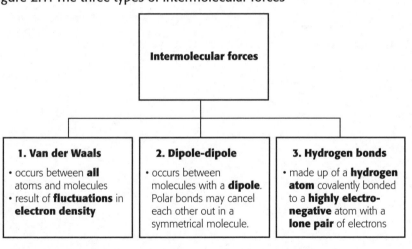

# Boiling points

$H_2O$, HF and $NH_3$ are all highly polar molecules. More energy is needed to overcome the additional hydrogen bonding between molecules. This makes their boiling points higher than predicted, as shown in the graph below:

**Figure 2.2: The boiling points of different hydrides**

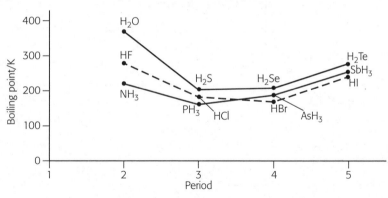

**KEY FACT**

Hydrogen bonding between water molecules means temperatures on Earth are not high enough for oceans to reach boiling point. Water molecules spread out when forming ice in order to maximize hydrogen bonding. This makes the ice less dense than liquid water, explaining why icebergs float.

**Bonding in the molecules of life**

✳ The two strands of the **DNA double helix** are held together by hydrogen bonds. They are strong enough to keep it together but weak enough to allow the strands to separate, allowing replication to take place.

✳ The tertiary (3D) structure of a protein is created by hydrogen bonding between different parts of the protein molecule.

# 3 Elements

# Introducing the elements

All materials are made from a combination of about 100 pure **elements**. Many elements are familiar, for example metals such as iron and copper and the non-metals, oxygen and helium.

Elements can combine to form **compounds**. Scientists need to be able to **predict** how the different elements and their compounds will react.

*Who dreamt up a way of organizing the elements that is still used today?*

The **periodic table** is a chart showing all the elements and allows scientists to make predictions about the properties and reactions of each element. It was first created by the Russian scientist **Dimitri Mendeleyev** – the idea for the table came to him in a dream – who organized the elements in order of mass.

This chapter focuses on four important groups of elements:

* **alkali metals**
* **alkaline earth metals**
* **halogens** *and*
* **noble gases**.

**The periodic table is a chart showing all the elements and allows scientists to make predictions about the properties and reactions of each element**

# The periodic table

The modern periodic table – see the one at the end of Chapter 10 – lists all the elements in order of **atomic number**.

Families of elements with similar properties belong to the same numbered vertical **group**:

* on the left-hand side are found the metals including group 1 (**alkali metals**) and group 2 (**alkaline earth metals**)
* on the far right-hand side are non-metal group 7, the **halogens**, and group 8, the unreactive **noble gases**.

The pattern, or trend, in the way one family reacts allows chemists to make predictions about reactions.

## Reactivity

The reactivity of group 1 (shown below) and group 2 **decreases** going up the group:

The reactivity of group 7 **increases** going up the group:

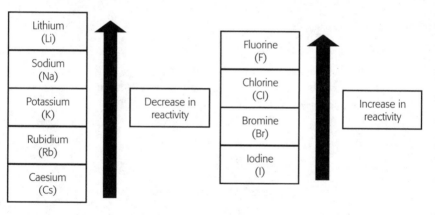

# The periodic groups

## Group 1

The alkali metals are unusual metals in several ways. Compared with most metals, an alkali metal is:

* lower in melting point
* less dense
* softer (Li, Na and K can be cut with a knife)
* more highly reactive (they react with water to produce an alkaline hydroxide solution).

> The alkali metals are so reactive that they must be stored in oil to prevent them reacting with air or water.

# Group 2

The alkaline earth metals consist of beryllium, magnesium, calcium, strontium and barium. The alkaline metals have:

* the usual properties of metals
* high melting points (unlike group 1)
* two electrons in the outer shell.

Melting points decrease going down the group because…

* the size of atoms increases
* the delocalized electrons are further from the nucleus so the strength of the metallic bonding decreases.

### DID YOU KNOW?
The bright white sparks from a firework may be created as highly reactive magnesium burns in the air.

# The periodic groups/displacement

## Group 7

The elements that make up group 7 are known as the halogens:

* their reactivity increases going up the group
* melting and boiling points increase going down the group – this is because there are more electrons so the van der Waals attraction is greater
* the halogens exist in different states at room temperature – fluorine and chlorine are gases, bromine is a liquid and iodine is a solid.

> **DID YOU KNOW?**
> Hydrofluoric acid is so corrosive that it cannot be kept in a glass bottle. Polyethylene or Teflon bottles must be used instead.

## Displacement reactions

When the halogens react, they gain one electron to become halide ions (e.g. $Cl^-$). A more reactive halogen will **displace** a less reactive one from solution.

For example, if chlorine solution is added to colourless potassium bromide solution, this solution gradually turns yellow. This is because chlorine has **displaced** bromine from the solution:

$$Cl_2 \text{ (aq)} + 2NaBr \text{ (aq)} \rightarrow Br_2 \text{ (aq)} + 2NaCl \text{ (aq)}$$

NB: Here (aq) indicates **aqueous** solution.

When writing a displacement reaction equation, check that:

* the halogens have the formula $F_2$, $Cl_2$, $Br_2$ or $I_2$ because they are **diatomic molecules** *and*
* the number of atoms is **balanced** on both sides of the equation.

# Patterns in reactivity

Patterns in the reactivity of the halide ions allow silver nitrate to be used to test which type of ion is present. First, silver nitrate solution is added and then the colour of the precipitate noted. The precipitates can be distinguished more definitively by further addition of ammonia solution.

## Table 3.1: Results from tests using silver nitrate

| Halide ion | With silver nitrate solution | With ammonia |
| --- | --- | --- |
| Fluoride | No ppt | – |
| Chloride | White ppt | Dissolves in dilute ammonia |
| Bromide | Cream ppt | Dissolves in concentrated ammonia |
| Iodide | Pale yellow ppt | Insoluble in concentrated ammonia |

NB: In chemistry, 'ppt' is the standard abbreviation of precipitate, the insoluble compound formed by the reaction of two solutions.

## CASE STUDY: A missing element

When creating the periodic table, Mendeleyev left gaps for those elements yet to be discovered and correctly predicted their properties.

Germanium was one of these missing elements. Germanium has properties in between silicon and tin, the previously discovered elements immediately above and below it in group 4. It is shiny, like a metal, but brittle, like a non-metal.

The element is now used to make semiconductors in electrical devices and in the optics of infrared systems such as night-vision goggles.

# 4 Trends

# Patterns in chemistry

It is always useful to understand the patterns found within different groups of the periodic table. But what of the transition metals between groups 2 and 3?

These transition metals are essential to many areas of industry. An understanding of the gradual change in properties of elements across the periodic table allows scientists and engineers to select the most appropriate materials.

*Why is titanium ideal for making aircraft?*

# The very name 'periodic table' reflects the periodic nature of the properties of the elements

The very name 'periodic table' reflects the periodic nature of the properties of the elements. This means that while vertical columns contain elements that are **similar**, the properties of the elements must change in a **gradual** fashion across each row, or **period**, of the table.

This chapter examines **trends** in properties such as:

* **atomic radius**
* **melting point** *and*
* **ionization energy**.

# Identifying trends

The summary in the following table gives the structure of each element in period 3 and whether it is a metal or non-metal.

## What trends can be seen?

**Table 4.1: Trends across period 3**

| Group | Element | Metal or non-metal | Structure |
|-------|---------|--------------------|-----------|
| 1 | Sodium | Metal | Giant metallic |
| 2 | Magnesium | Metal | Giant metallic |
| 3 | Aluminium | Metal | Giant metallic |
| 4 | Silicon | Semi-metal | Giant covalent |
| 5 | Phosphorus | Non-metal | Simple molecular |
| 6 | Sulphur | Non-metal | Simple molecular |
| 7 | Chlorine | Non-metal | Simple molecular |
| 8 | Argon | Non-metal | Simple molecular |

The melting point of the metals and silicon are **high** as they have giant structures with **strong metallic or covalent bonds within** the structure. The melting point of the non-metals is much **lower** as only **relatively weak van der Waals forces** act **between** molecules.

**Figure 4.1: The melting points of elements in group 3**

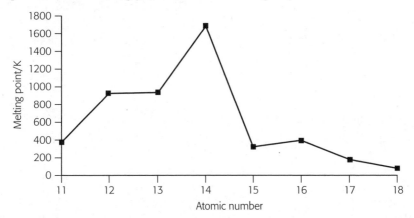

# Atomic radius

## ...atomic radius is taken as half the distance between the nuclei of two covalently bonded atoms

The definition of **atomic radius** is not straightforward because there is no defined point at which the electron density drops to zero. For this reason, atomic radius is taken as half the distance between the nuclei of two covalently bonded atoms.

Look at the atomic radii of period 3 (all units are in nm (nanometres)):

Na 0.156, Mg 0.136, Al 0.125, Si 0.117, P 0.110, S 0.104, Cl 0.099

*What trend can you see in the atomic radius?*

We know that:

* across period 3, electrons are gradually added to the outer third shell
* each additional electron is roughly the **same distance** from the nucleus and at the same time additional protons are added to the nucleus
* **positive charge increases** from +11 to +17 and this causes **increased electrostatic attraction** with the electrons
* the electrons are **pulled more closely** towards the nucleus
* the **atomic radii decrease** across the period.

### Transition metals

The transition metals are found between groups 2 and 3. The decrease in atomic radius is relatively small compared with the increase in mass of the nucleus. For this reason, titanium is much less dense (density = mass divided by volume) than iron, making it ideal for use in aircraft.

# Ionization energy

In order to form ions, outer electrons must escape the pull of the positive nucleus.

The energy required to remove the first electron from 1 mole of atoms is known as the **first ionization energy** and is defined as:

**'The energy required to remove 1 mole of electrons from 1 mole of gaseous atoms.'**

44

In general, more energy is required to remove electrons from shells closer to the nucleus because electrostatic attraction increases with decreasing distance, so first ionization energy increases moving up a group. First ionization energy also increases as you move across a period.

Any explanation for this should mention:

* the electron shell in which the electron is located
* the electron's distance from the nucleus
* the charge of the nucleus
* the strength of the electrostatic attraction between the electron and the nucleus.

## ...more energy is required to remove electrons from shells closer to the nucleus...

# Ionization energy: exceptions

A more detailed model of electron arrangements uses **subshells**. Electron subshells can help explain those exceptions where the first ionization energy does not increase as expected.

For example, aluminium ($1s^2$, $2s^2$, $2p^6$, $3s^2$, $3p^1$) has a lower ionization energy than magnesium ($1s^2$, $2s^2$, $2p^6$, $3s^2$). This is because the 'p' subshell is at a slightly higher energy level and so less energy is needed to remove this electron than one in the 's' subshell.

> **TOP TIP**
> To work out the electron arrangement of an atom, fill up the sublevels in order (the maximum number of electrons is given in square brackets):
> 1s[2], 2s[2], 2p[6], 3s[2], 3p[6]

Table 4.2: Electron arrangements for periods 1 and 2

| Element | Number of electrons | 1s | 2s | 2p | | |
|---------|---------------------|-----|-----|-----|-----|-----|
| H | 1 | ↑ | | | | |
| He | 2 | ↑↓ | | | | |
| Li | 3 | ↑↓ | ↑ | | | |
| Be | 4 | ↑↓ | ↑↓ | | | |
| B | 5 | ↑↓ | ↑↓ | ↑ | | |
| C | 6 | ↑↓ | ↑↓ | ↑ | ↑ | |
| N | 7 | ↑↓ | ↑↓ | ↑ | ↑ | ↑ |
| O | 8 | ↑↓ | ↑↓ | ↑↓ | ↑ | ↑ |
| F | 9 | ↑↓ | ↑↓ | ↑↓ | ↑↓ | ↑ |
| Ne | 10 | ↑↓ | ↑↓ | ↑↓ | ↑↓ | ↑↓ |

# 5 Energy

## Understanding energy changes

'Calorie counting' is a classic form of dieting in which energy intake is minimized in order to lose weight. These days the standard unit of energy is the **joule** with the energy content of food being given in **kilojoules**.

The energy content of food is worked out by measuring the energy released when the food is burnt. Simple equipment may be used in the lab to measure the approximate energy change of many other types of chemical reaction.

*Why and how is it possible to mix chemicals inside a cool pack to make it turn cold?*

## ...the standard unit of energy is the joule with the energy content of food being given in kilojoules

This chapter addresses the fundamentals of understanding the **energy changes** that take place when different chemicals react. It considers:

* the energy changes for particular processes (**combustion** and **formation** of elements) *as well as*
* the energy changes as **bonds** within the chemical **break** or **form**.

Importantly, the term '**enthalpy change**' is introduced in terms of a change in heat at constant pressure.

# Exothermic or endothermic?

* An **exothermic** reaction transfers energy **out** to the surroundings. During the burning of fuels heat is transferred to the surroundings, hence **combustion** reactions are exothermic.

* An **endothermic** reaction takes in energy from the surroundings. If you heat calcium carbonate, it decomposes to form calcium oxide and carbon dioxide gas. **Thermal decomposition** is therefore an endothermic reaction.

52

### TOP TIP

When considering heat transfer in reactions, remember: the **ex**it is the way out; the **en**trance is the way in. So, during an exothermic reaction, heat **ex**its to the surroundings; during an endothermic reaction, heat **en**ters from the surroundings.

## Making and breaking bonds

When methane gas burns, the bonds in the methane and oxygen molecules must be **broken**. This requires energy *from* the surroundings. When the bonds in the carbon dioxide and water molecules are formed, energy is released *to* the surroundings. The energy released is greater than the energy needed, so overall the reaction is exothermic:

$$CH_4 + 2O_2 \rightarrow CO_2 + 2H_2O \ (-276 \text{ kJ mol}^{-1})$$

Exothermic reactions always have a **negative energy change**. This is because the products have **less energy** than the reactants as energy has been transferred to the surroundings.

In an endothermic reaction the reactants *take in* energy from the surroundings. By the end of the reaction the products have **more energy** than the reactants so the energy change is **positive**.

## Enthalpy change

The term 'enthalpy change' is used to describe the energy change of a reaction at constant pressure.

Enthalpy changes are usually given under standard conditions and are given the symbol $\Delta H_{298}^{\theta}$ where the Greek letter $\Delta$ represents change and H stands for entHalpy.

**Note!**
Standard temperature is 298 K and standard pressure is 100 kPa (or 1 atm).

Some enthalpy changes have specific definitions. For example:

## 'The standard molar enthalpy of formation $\Delta H_f^{\theta}$ is the enthalpy change when one mole of a compound is formed from its elements in their standard states under standard conditions.'

## Hess's Law

According to Hess's Law:

> ## 'The enthalpy change for a chemical reaction is the same, whatever route is taken from reactants to products.'

When drawing a Hess's Law diagram…

✔ Do start by writing out the reaction for which you are trying to find the enthalpy change

✔ Do identify an enthalpy change that could make the molecules in your reaction (usually $\Delta H_f^{\circ}$)

✔ Do use the correct sign + or − for the enthalpy changes

✘ Don't forget to check the direction of the reaction arrows

✘ Don't forget to multiply the enthalpy changes where more than one mole of molecules is being changed

# Measuring enthalpy change

If a food sample is burnt underneath a copper can of water, the change in temperature can be recorded and the enthalpy change calculated using the following equation:

Enthalpy change (J) = mass of substance (g) ×
specific heat capacity ($Jg^{-1} K^{-1}$) × temperature change (K)

A **bomb calorimeter** is used in professional laboratories as it is designed to minimize any heat loss to the surroundings and to reduce other sources of error.

## 'The standard molar enthalpy of combustion, $\Delta H_c^{\theta}$, is the enthalpy change when one mole of a compound is burnt completely in oxygen under standard conditions.'

Measurement of enthalpy changes during a reaction in solution requires – at a minimum – a polystyrene cup (with a lid) and a thermometer.

By keeping any heat released by the reaction inside the solution, the temperature increase can be measured and the enthalpy change calculated.

### Note!
Never reuse a polystyrene cup used in a lab. Traces of harmful chemicals may remain.

# Calculating enthalpy change

Theoretical calculations may be carried out using known values of $\Delta H_f^{\theta}$. There are two methods:

1. Using a **thermochemical cycle**. The reactants and products make two corners of a triangle and the elements from which they are made make a third.

2. Using an **enthalpy level diagram**. The elements should be at energy level 0, with the reactants and products placed appropriately according to the $\Delta H_f^{\theta}$.

On an enthalpy level diagram, a **downwards** arrow represents an **exothermic** reaction (**negative $\Delta H$**).

## Figure 5.1: An enthalpy level diagram

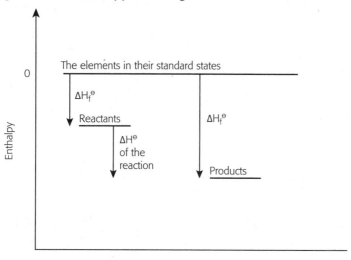

To find the enthalpy change of the reaction find the difference between levels of the reactants and products.

# 6 Rates

# The rate of reactions

An iron bridge can gradually rust. If properly maintained and painted, the rust will not occur at all. The paint layer stops the reacting oxygen molecules from making contact with iron atoms. No reaction can occur unless the reacting particles can meet and collide.

In contrast, a strip of magnesium ribbon will burn brightly in air. The rate of this reaction is much faster. Chemists need to be able to control the **rate of reactions**.

*How is the conversion of a car's exhaust gases speeded up?*

Understanding **rate of reaction** and **equilibrium** is critical for the chemical industry. Products must be made quickly and with maximum yield.

This chapter outlines:

* how to increase the rate of reaction
* what is happening at the **molecular level** during reactions
* how **catalysts** work *and*
* how conditions can affect the **equilibrium position** between forward and reverse reactions.

## Understanding rate of reaction and equilibrium is critical for the chemical industry

# Collision theory

A chemical reaction cannot take place unless the participating atoms or molecules **collide**. In order for a reaction to take place, the molecules must have **sufficient kinetic energy** when they collide. The molecules must also have the right **orientation**.

The higher the proportion of collisions that are successful, the faster the reaction. This can be achieved either by increasing the number of molecules in a given volume of space or by increasing the kinetic energy of the molecules.

Chemists can also use catalysts to speed up a reaction.

Rate of reaction can be increased by:

1 **increasing the temperature** – this increases the kinetic energy of the molecules so the number of collisions will increase and these collisions will have more energy
2 **increasing the pressure of gases** – an increase in the number of particles in a given volume means that collisions will happen more frequently
3 **increasing the concentration of the solution** – again with more molecules in the same volume, collisions will happen more frequently
4 **increasing the surface area of any solid reactants** – a larger surface area means that more atoms are exposed to the other reactants and more collisions will occur.

### Key term: catalysts
Catalysts are additional chemicals that increase the rate of reaction without being chemically changed themselves.

# Activation energy

The minimum energy needed for a reaction to start is termed the **'activation energy'**. On an enthalpy level diagram this can be pictured as the energy increase needed to reach a **transition state**. Once this energy level has been reached the reaction can proceed.

* For an **exothermic** reaction, the resulting **products** will be at a **lower energy level** than the reactants.
* For an **endothermic** reaction, the **products** will have a **higher energy level** than the reactants.

## Maxwell-Boltzman distribution

The molecules in a gas each have different amounts of kinetic energy. A few molecules have very little kinetic energy and a few have a lot.

The range of energies is shown in the **Maxwell-Boltzman distribution**. By increasing the temperature, the number of molecules with energy above the activation energy level increases.

**Figure 6.1:** The Maxwell-Boltzman distribution for molecules at two temperatures

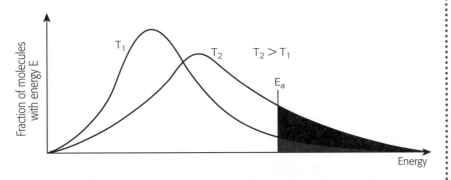

The grey shaded area represents that number of molecules with energy greater than the activation energy $E_a$. Note that when the temperature is increased to $T_2$, the area under the curve (shaded black) increases. This means that more molecules will have sufficient energy to react when they collide. The rate of reaction is therefore faster.

# Catalysts

Catalysts work by **decreasing the activation energy**. On the Maxwell-Boltzman distribution this may be pictured by moving the vertical activation energy line to the left.

The area under the graph to the right of this increases, meaning that a greater fraction of molecules have sufficient energy to react.

## Catalysts work by decreasing the activation energy

**Note!**
Catalysts use **ad**sorption NOT **ab**sorption (as with a sponge). The greater the surface area of the catalyst, the more effective it will be.

## CASE STUDY: Catalytic converters

Modern petrol-engined cars are fitted with catalytic converters. These speed up the conversion of polluting exhaust gases into less polluting alternatives. This is how they work:

1 The polluting gases form weak bonds with metal atoms coating the ceramic catalyst.
2 This process is called **adsorption** and is maximized by the honeycomb structure of the catalytic converter.
3 The reactions occur faster on the surface of the catalyst as the molecules are held in the optimum orientation.
4 The products are released from the catalyst in a process called **desorption**.
5 The catalyst is unchanged and more molecules may now react.

# Equilibrium in reactions

During a reaction not all reactants are necessarily converted to products. The balance or **equilibrium** between the forward and reverse reactions depends on the relative rate of each reaction. For example:

* during the **Haber process**, nitrogen and hydrogen gas react to form ammonia
* in the reverse reaction, ammonia may break down into the constituent gases
* the **equilibrium position** depends on the conditions of the reaction which affect the rate of the forward and reverse reactions

$$N_2\ (g) + 3H_2\ (g) \rightleftharpoons 2NH_3\ (g)\ (\Delta H^\theta = -92\ kJ\ mol^{-1})$$

NB: Here (g) indicates gas.

## The balance or equilibrium... depends on the relative rate of each reaction

## Le Châtalier's principle

Le Châtalier's principle states:

> **'In an equilibrium system any change to the system will move the equilibrium in a direction that will counteract that change.'**

During the Haber process an increase in pressure will move the equilibrium position to the right. This reduces the number of gas molecules and will therefore reduce the pressure again.

An increase in temperature will move the equilibrium to the left. This results in more energy being taken from the surroundings, thereby lowering the temperature again.

# 7 Quantitative chemistry

# Introducing quantitative chemistry

Numbers help chemists in all sorts of ways. Chemists need numbers to discuss how many molecules are reacting. They also need them to be able to calculate the concentration of a solution.

What's more, balancing equations can become increasingly tricky, so again numbers, in the form of oxidation states and transferring electrons, make the job easier.

*How could you prove whether some vinegar had been watered down?*

● Quantitative chemistry is easier to understand than you think.

Quantitative chemistry can be a cause of stress, but it need not be.

This chapter covers two essential areas:

* **moles** – these are essential to understand in order to work out the concentration of a particular solution
* **redox reactions** – these show both oxidation and reduction. An understanding of electron transfer and oxidation numbers helps to balance this type of equation.

Quantitative chemistry is also essential for analysis of solutions using titration techniques. In this, a solution of known concentration is reacted quantitatively with a solution of unknown concentration. This chapter summarizes the basic method for using information from a titration to calculate the unknown concentration.

# Atomic mass

It would be impractical for scientists to refer constantly to the mass of single atoms. The numbers are far too small to deal with. Instead, scientists consider relative atomic masses in comparison with 12 g of carbon-12.

The relative atomic mass of an atom is the ratio of the average mass of an atom of an element (of the natural isotopic composition) to 1/12th the relative atomic mass of an atom of carbon-12.

The number of atoms in 12 g of carbon-12 is called the **Avogadro constant**. It is usually written as $6.02 \times 10^{23}$. The amount of a substance that contains this many individual entities (atoms, molecules, ions…) is called a **mole**.

For compounds, relative formula mass is used. This may be calculated by adding together the relative atomic masses of all the atoms in the formula.

**KEY FACT**
The number of atoms in 12 g of carbon-12 is
602,200,000,000,000,000,000,000.

**Worked example**
What is the relative formula mass of NaCl
(sodium chloride) and $H_2SO_4$ (sulphuric acid)?

$NaCl = 23 + 35.5 = 58.5$
$H_2SO_4 = (2 \times 1) + 32 + (4 \times 16) = 98$

# Concentration

The concentration of a solution is measured in moles per decimetre cubed ($mol\ dm^{-3}$). To calculate the concentration of a solution you need to know the number of moles of the solute and the volume of the solution in $dm^3$.

* Number of moles = mass in g ÷ mass of 1 mol in g
* Concentration = number of moles ÷ volume in $dm^3$

**TOP TIP**

Always make sure you measure the volume in $dm^3$ and not $cm^3$. To change $cm^3$ to $dm^3$, divide by 1,000.
For example:
$25\ cm^3 = 0.025\ dm^3$ (calculated by dividing 25 by 1,000)

The concentration of a solution can be checked or worked out by titrating it with a solution of known concentration. Here's an example of a titration calculation:

It takes 20.0 cm$^3$ of hydrochloric acid (HCl) to neutralize 25.0 cm$^3$ of 0.100 mol dm$^{-3}$ sodium hydroxide (NaOH) solution. What is the concentration of the acid?

1  Write down the equation of the reaction:

$$HCl\ (aq) + NaOH\ (aq) \rightarrow NaCl\ (aq) + H_2O\ (l)$$

2  Work out how many moles of NaOH have reacted:

= concentration × volume (of NaOH)
= 0.100 × 0.025 dm$^3$ = 0.0025 moles

3  Use the 1:1 ratio in the equation to infer that the same number of moles of HCl have reacted.

4  Then input numbers into the equation for concentration:

= moles HCl ÷ volume HCl
= 0.0025 ÷ 0.020 = 0.125 mol dm$^{-3}$

# Redox reactions

When magnesium (Mg) burns in oxygen ($O_2$), producing magnesium oxide (MgO), we say it has been oxidized. Magnesium has lost electrons to become a magnesium ion ($Mg^{2+}$) so **oxidation** may be described as the **loss of electrons**.

At the same time oxygen **has gained electrons** to become $O^{2-}$. Oxygen has been **reduced**.

A reaction in which both reduction and oxidation occur is known as a **redox reaction**.

### TOP TIP
Remember the acronym OILRIG: Oxidation Is Loss Reduction Is Gain (of electrons).

# Half equations

Redox reactions are described using **half equations**. These show clearly how the electrons are being transferred. These must be balanced so that the number of electrons gained is the same as the number of electrons lost.

The half equations for the burning of magnesium are:

$$Mg \rightarrow Mg^{2+} + 2e^- \text{ and } \tfrac{1}{2}O_2 + 2e^- \rightarrow O^{2-}$$

* An oxidizing agent **removes** electrons.
* A reducing agent **gives** electrons.

In this case magnesium is the reducing agent and oxygen is the oxidizing agent.

The half equations can then be combined to form a complete equation:

$$\overset{-2e^-}{\overbrace{Mg + \tfrac{1}{2}O_2 \rightarrow MgO}}$$
$$\underset{+2e^-}{}$$

# Oxidation states

In order to balance equations including non-ionic compounds (such as $H_2O$) a system of **oxidation states** (or numbers) is used.

Oxidation numbers are assigned according to the following rules:

1 Atoms of any element have oxidation state 0.
2 The oxidation state of a simple ion is the same as the charge of the ion.
3 The oxidation states of a compound must add up to 0.
4 The oxidation states of a complex ion must add up to the charge of the ion.
5 Some elements always have the same oxidation number and so may be used to derive the oxidation state of other elements in a compound.

It is important to remember those elements which usually have the same oxidation state and also the exceptions.

**Table 7.1: Oxidation states of different elements**

| Element(s) | Oxidation state |
| --- | --- |
| Group 1 | +1 |
| Group 2 | +2 |
| Aluminium | +3 |
| Fluorine | −1 |
| Oxygen | −2 (except in peroxides $O_2^{2-}$ or in compounds with F where it is −1) |
| Hydrogen | +1 (except in metal hydrides $H^-$) |
| Chlorine | −1 (except when combined with O or F) |

What is the oxidation state of iron in $Fe_2O_3$? The individual oxidation state of oxygen is −2, so the combined oxidation states of oxygen add up to −6. Therefore the combined oxidation states of iron must add up to +6. Each iron ion must have an oxidation state of +3.

# 8 Organic molecules

# What are organic molecules?

An organic molecule is based on a simple **hydrocarbon chain**: $CH_3[CH_2]_nCH_3$. The chain may be **branched** or contain **double** or **triple bonds**.

Other atoms, such as oxygen or nitrogen, may be attached to the main hydrocarbon chain as part of a distinctive **functional group**.

This chapter introduces how scientists categorize and name organic molecules in order to help predict their properties and reactions.

*On a very cold day, a diesel car won't start. Why? Are petrol engines affected in the same way?*

## ...scientists categorize and name organic molecules in order to help predict their properties and reactions

••••••••••••••••••••••••••••••••••••••••••••••••••

Without some sort of organization, chemists would need volumes to describe all organic molecules. In order to communicate unambiguously, scientists have developed a set of rules governing their **nomenclature** (naming).

This chapter gives some basic rules on naming organic molecules. It also summarizes how these are represented by **formulae** and looks at patterns in the properties of the most simple family of molecules – the **alkanes**.

# Naming carbon chains

The root of all organic molecule names is based on the number of carbons in the main chain. For example: 1 = meth, 2 = eth, 3 = prop, 4 = but, 5 = pent, 6 = hex.

The ending **'ane'** means the molecule consists of all single bonds while the ending **'ene'** indicates the presence of double bonds.

Additional branches have prefixes based again on the number of carbon atoms. For example: 1 = methyl, 2 = ethyl, 3 = propyl. A number is also used to indicate the carbon in the main chain to which the branch is attached.

## The root of all organic molecule names is based on the number of carbons in the main chain

# Formulae

Organic molecules may be described by four different types of formula:

1 **empirical** formula – this gives the simplest ratio of atoms in the molecule
2 **molecular** formula – this gives the actual number of each type of atom
3 **structural** formula – this shows how atoms are grouped within the molecule
4 **displayed** formula – this shows both the atoms and the bond (single or double or triple) within the molecule.

For example, the different formulae for butane (4 carbon chain) are:

* empirical formula: $C_2H_5$
* molecular formula: $C_4H_{10}$
* structural formula: $CH_3CH_2CH_2CH_3$
* displayed formula:

# Functional groups

The name of an organic molecule is based on the **functional group** and the **length of the main carbon chain**. Ethanol, for example, is an alcohol with two carbons in the main chain.

### Table 8.1: Functional groups of organic molecules

| Family | Functional group | Example with 3 carbon chain |
|---|---|---|
| Alk**ane**s | None | Prop**ane** |
| Alk**ene**s | R–CH=CH$_2$ | Prop**ene** |
| Halogenoalkanes | R–X where X is a halogen (F, Cl, Br or I) | **Bromo**propane (or chloro, fluoro or iodo) |
| Alcoh**ol**s | R–OH | Propan**ol** |
| Aldehydes | R–CHO | Propan**al** |
| Ketones | R–CO–R' | Propan**one** |
| Carboxylic acid | R–COOH | Propan**oic acid** |

NB: Here R represents any length of basic hydrocarbon chain. R' is used to indicate a potentially different hydrocarbon chain to R.

## Isomers

A 5 carbon chain may be arranged in three different ways:

1 $CH_3CH_2CH_2CH_2CH_3$ (pentane)
2 $CH_3CH(CH_3)CH_2CH_3$ (2-methylbutane)
3 $CH_3C(CH_3)_2CH_3$ (2,2-dimethylpropane).

Molecules with the same atoms but different arrangements are called **isomers**.

### TOP TIP
Always check that an isomer really is a **different** molecule. You may need to imagine rotating the molecule. For example, $CH_3CH_2CH(CH_3)CH_3$ is *not* a new isomer. It is the same as the second example given above.

# Alkanes

* Each alkane is made of a simple **hydrocarbon chain**.
* Alkanes are **non-polar** and are **insoluble** in water.
* Alkane molecules are attracted to each other by weak **van der Waals** forces.
* Longer alkanes have more electrons so the van der Waals attraction is greater. Therefore **longer** alkanes have **higher boiling points**.
* Branching in the carbon chain reduces the van der Waals attraction and decreases the boiling point.

### DID YOU KNOW?
Crude oil is a mixture that is separated by **fractional distillation**. This relies on the difference in boiling points of different alkanes.

## CASE STUDY: Uses of alkanes

Alkanes…

✳ are used as **fuels** in cars, homes, aeroplanes and ships
✳ undergo **combustion** producing **carbon dioxide** and **water**
✳ may be **cracked** to form an alkene (used to make polymers) or
  shorter alkanes (which may be of more use as fuels).

Alkanes may undergo incomplete combustion, if there is insufficient oxygen, producing carbon monoxide. Carbon monoxide has tragically led to many deaths. This is why gas appliances must be regularly serviced.

### DID YOU KNOW?

As diesel molecules are longer than those of petrol, diesel starts to solidify in very cold conditions whereas petrol remains a liquid.

# Halogenoalkanes

The **halogenoalkanes** consist of a hydrocarbon chain with at least one additional halogen atom. Just one halogen atom is enough to change the properties of the molecule.

A halogen atom is **electronegative** and attracts electron density away from the neighbouring carbon atom. The carbon atom has a slight **positive** charge ($\partial^+$) and the halogen atom a slight **negative** charge ($\partial^-$).

The bond is described as **polar**. A **dipole** has been formed. This means that both van der Waals and dipole–dipole intermolecular forces are acting, so boiling points are **higher** than the alkanes.

NB: The Greek letter $\partial$ is used to indicate a small change.

# The reactivity of halogenoalkanes

✳ The halogens get more electronegative as you go up the group.

✳ This makes a carbon atom next to a fluorine atom more $\partial^+$ than if next to an iodine atom. But, this does *not* make a fluoroalkane more reactive.

✳ A fluorine atom is small with electrons close to the nucleus.

✳ The covalent bond with the neighbouring carbon atom is very strong. This gives a very high bond enthalpy so the bond needs a lot of energy to break.

✳ The reactivity of the halogenoalkanes **decreases going up** group 7.

**Note!**
The reactivity of the halogenoalkanes is opposite to the reactivities of the halogen group.

# 9 Mechanisms

## The mechanism of reactions

Some organic molecules have been used for hundreds of years. Queen Victoria used the halogenoalkane, chloroform, to ease the pains of childbirth. In the twentieth century scientists learnt to design organic molecules to suit a specific purpose, e.g. nylon for stockings or indigo dye for jeans. These molecules could then be manufactured and sold to consumers.

Gaining an understanding of how different organic molecules react allows scientists to predict the properties of products and/or amend reaction conditions to ensure a good yield of the final product.

*Which alkene is needed to create the protective layer that stops bacon sticking to a frying pan?*

To work out how to make a particular molecule from a given reactant, scientists consider *how* the reaction could take place. This is known as the **mechanism** of the reaction. This depends on the starting molecule and how it interacts with a particular type of reagent.

This chapter considers three main mechanisms of reaction...

* **nucleophilic substitution**
* **elimination** *and*
* **electrophilic addition**

...and how they apply to the **halogenoalkanes** and **alkenes**.

# Reactions of halogenoalkanes

The slightly positive carbon atom neighbouring the halogen atom attracts electron-rich reagents (**nucleophiles**). A nucleophile can then replace the halogen atom in the molecule. Halogenoalkanes therefore undergo **nucleophilic substitution** reactions.

## A nucleophile is an ion or molecule capable of donating a lone pair of electrons

**TOP TIP**

Remember:

* bibliophile = lover of books
* nucleophile = lover of nucleus (i.e. positive charge)
* electrophile = lover of electrons

By reacting a halogenoalkane with the right nucleophile, it is possible to synthesize a range of different molecules.

**Figure 9.1: Reactions between a halogenoalkane and nucleophiles**

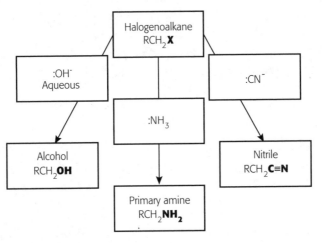

# Chain reactions

Halogenoalkanes may be formed from alkanes by the three stages of a **chain reaction**:

1 **Initiation** – Energy from ultraviolet (UV) light splits a chlorine molecule producing two chlorine **free radicals**.
2 **Propagation** – The chlorine free radicals react further, creating methyl free radicals, which in turn react with more chlorine molecules making yet more chlorine free radicals. The reaction is self-propagating.
3 **Termination** – The reaction can only stop if the free radicals are removed by reacting with each other.

UV light splits the Cl–Cl bond evenly, or homolytically, with one electron going to each chlorine atom. These free radicals are highly reactive and are written like this: Cl•

## CASE STUDY: Ozone depletion

Free radicals in the upper atmosphere have had the catastrophic effect of gradually destroying the ozone layer.

CFCs (chlorofluorocarbons) were widely used as refrigerants, solvents and as propellants in aerosols. Their unusually long life allowed them to spread into the upper stratosphere. Here, exposure to UV light broke the C–Cl bonds within the molecules. The resulting free radicals initiated a chain reaction that attacked ozone ($O_3$).

The Montreal Protocol, which came into force in 1989, called for a reduction in the use of these compounds. Today they have been largely replaced by alternative, safer compounds including hydrochlorofluorocarbons (HCFCs). Unfortunately there are still many CFCs remaining in the atmosphere so it will take many years for the ozone layer to fully recover.

# Reaction conditions

The **conditions** for an organic reaction are critical. For example, when reacting with a halogenoalkane, cold OH⁻ in aqueous solution (water) favours the production of an alcohol via nucleophilic substitution.

However, replace this with hot OH⁻ in ethanol and the favoured result is an alkene. In this case, the OH⁻ eliminates both H and the halogen atom (X) from the molecule, resulting in the formation of a double bond. Hence this type of reaction is called an **elimination reaction**.

## Alkenes

The family of organic molecules containing at least one carbon–carbon double bond are known as the **alkenes**. The name of an individual molecule is based on the longest carbon chain that contains the double bond. A number is inserted to indicate the position of the double bond.

The alkene with the molecular formula $C_4H_8$ has two isomers based on the 4 carbon chain:

$CH_2$=CH–$CH_2CH_3$ (but-1-ene) *and* $CH_3$CH=CH$CH_3$ (but-2-ene)

Look out for **geometric isomers**. The double bond cannot rotate so…

Z-but-2-ene          E-but-2-ene

…are different molecules.

The isomer with both $CH_3$ groups on the same side is known as Z-but-2-ene (from the German *Zusammen*, meaning together), whereas the molecule with $CH_3$ groups on opposite sides is called E-but-2-ene (from the German *Entgegen*, meaning opposite).

# Electrophilic addition

Here is an example of **electrophilic addition**:

1 HCl is a polar molecule so the H atom has a slight positive charge ($H^{\partial+}$).
2 $H^{\partial+}$ – called an **electrophile** – is attracted to the electron-rich double bond of ethene.
3 One electron from the double bond creates a bond with H.
4 The other electron from the double bond creates a Cl⁻ ion.
5 The Cl⁻ ion is attracted to the positive charge and is also **added** to the molecule.
6 The product 1-chloroethane is formed.

Electrophilic addition occurs via a temporary intermediate:

$$CH_2CH_2 + HCl \rightarrow CH_3CH_2{}^+ + Cl^- \rightarrow CH_3CH_2Cl$$

## Polymers

The alkenes are essential in the synthesis of **polymers**. The double bond is able to 'open' to link lots of alkene molecules (monomers) to make a long chain.

### Table 9.1: Examples of polymers and their monomers

| Common name of polymer | Systematic name | Monomer |
|---|---|---|
| Polythene | Poly(ethene) | $CH_2{=}CH_2$ |
| PVC (polyvinylchloride) | Poly(chloroethene) | $CHCl{=}CH_2$ |
| Teflon | Poly(1,1,2,2-tetrafluoroethene) | $CF_2{=}CF_2$ |

Polythene is widely used for products ranging from shopping bags to washing-up bowls. PVC's properties make it ideal for guttering and window frames, whereas Teflon can be used to give a non-stick lining to cooking equipment.

# 10 Synthesis

# Synthesizing molecules

## ...molecules can be designed and then synthesized for a very specific purpose...

Ethanol (alcohol) is produced by the natural process of fermentation. However, with an in-depth understanding of reactions, chemists can plan a step-by-step route to synthesize a molecule in the lab. In other cases, molecules can be designed and then synthesized for a very specific purpose, such as in the pharmaceutical industry.

*What mechanism enabled the manufacture of nylon stockings?*

Many items from everyday life, from bin liners to medicines, have been produced as a result of **organic synthesis**.

Starting molecules are transformed via a series of reactions into practical materials. Alcohol can be the starting point for a variety of such useful reactions.

This chapter outlines basic **reactions of alcohols** that – together with reactions outlined in the previous chapters – may be used to design the synthesis of some organic molecules.

# Alcohols

All alcohols contain one or more of the OH functional groups. They are named according to the longest carbon chain and the position of the OH group(s). For example:

$$CH_2OHCH_2CH_3 \text{ is propan-1-ol } whereas$$

$$CH_3CH(OH)CH_3 \text{ is propan-2-ol}$$

Alcohols are classified according to the carbon to which OH is attached (shown in bold above):

* in a **primary** alcohol, this carbon atom is attached to one other
* in a **secondary** alcohol, this carbon is attached to two others
* in a **tertiary** alcohol, this carbon is attached to three others.

NB: Propan-1-ol is a primary alcohol and propan-2-ol is a secondary alcohol.

The position of the OH group is critical in determining the products of reaction.

> **TOP TIP**
> Remember how alcohols are classified by thinking of the order of education establishments:
>
> * primary school
> * secondary school
> * tertiary education

## Experimental techniques

Two key processes that are routinely used when reacting alcohols are:

1 **Distillation** When the reacting mixture is heated, the desired product forms a vapour. This rises out of the reacting vessel and enters the condenser. Here it is cooled and collected.
2 **Reflux** Where the reactants need prolonged heating, the condenser is arranged directly above the reacting vessel so that any intermediary products drip back into the reacting mixture for further heating.

# Aldehydes and ketones

Alcohols may be oxidized to form two other types of organic molecule:

## Aldehydes

* **Aldehydes** have a C=O group at the end of the carbon chain.
* They are named after the length of the carbon chain and have the ending: **-al**.
* Aldehydes can be oxidized to form carboxylic acids.
* Aldehydes produce a layer of metallic silver on the inside of the test tube if reacted with **Tollen's reagent**.
* They are also oxidized by **Fehling's** or **Benedict's reagent** to form a brick-red precipitate.

> **DID YOU KNOW?**
> Common names of carboxylic acids include:
>
> * formic acid (produced by ants) – also known as methanoic acid
> * acetic acid (vinegar) – also known as ethanoic acid.

# Ketones

* **Ketones** also have a C=O group, which is within the hydrocarbon chain.
* Ketones are named after the length of the carbon chain and have the ending: **-one**.
* Ketones are not further oxidized and thus give a negative result when reacted with Tollen's, Fehling's or Benedict's reagents.

> **DID YOU KNOW?**
> The ketone commonly known as acetone (propanone) is the active ingredient in nail polish remover.

# Reactions: oxidation and elimination

Alcohols are very useful during the synthesis of new molecules as – depending on the type of alcohol and the conditions – different products may be created. First, alcohols can undergo **oxidation** reactions:

**Figure 10.1: Examples of oxidation reactions**

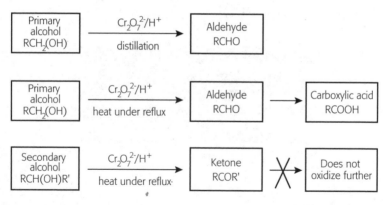

Alcohols can also undergo **elimination** reactions. As it is a water ($H_2O$) molecule that leaves, an alternative name for the process is **dehydration**.

The OH comes from the functional group but another H must leave the hydrocarbon chain to make the $H_2O$ molecule. This results in a C=C double bond. The product is an **alkene**.

# An elimination reaction is a reaction in which a small molecule leaves the main molecule

## DID YOU KNOW?

The polymer nylon is produced through an elimination reaction. As each monomer joins the polymer chain one molecule of $H_2O$ is released. For this reason this process is known as condensation polymerization. Nylon was discovered by scientists working for the Dupont company and first used to make stockings in 1939.

# Organic synthesis

When planning an organic synthesis you need to:

* note the arrangement of the hydrocarbon chain and any functional groups on the target molecule
* work backwards:
    » which molecules could react to form your target molecule?
    » which molecules could react to form those molecules?
* stop when you reach a suitable starting molecule
* note down the reagents and conditions needed for each step.

*Always carry out a risk assessment, or consult someone qualified to do so, before starting a chemical process.*

# The periodic table

| 1 | 2 | | | | | | | | | | | 3 | 4 | 5 | 6 | 7 | 8 |
|---|---|---|---|---|---|---|---|---|---|---|---|---|---|---|---|---|---|
| | | | | | | | | | | | | | | | | | He 2 |
| Li 3 | Be 4 | | | | | | | | | | | B 5 | C 6 | N 7 | O 8 | F 9 | Ne 10 |
| Na 11 | Mg 12 | | | | | | | | | | | Al 13 | Si 14 | P 15 | S 16 | Cl 17 | Ar 18 |
| K 19 | Ca 20 | Sc 21 | Ti 22 | V 23 | Cr 24 | Mn 25 | Fe 26 | Co 27 | Ni 28 | Cu 29 | Zn 30 | Ga 31 | Ge 32 | As 33 | Se 34 | Br 35 | Kr 36 |
| Rb 37 | Sr 38 | Y 39 | Zr 40 | Nb 41 | Mo 42 | Tc 43 | Ru 44 | Rh 45 | Pd 46 | Ag 47 | Cd 48 | In 49 | Sn 50 | Sb 51 | Te 52 | I 53 | Xe 54 |
| Cs 55 | Ba 56 | La* 57 | Hf 72 | Ta 73 | W 74 | Re 75 | Os 76 | Ir 77 | Pt 78 | Au 79 | Hg 80 | Tl 81 | Pb 82 | Bi 83 | Po 84 | At 85 | Rn 86 |
| Fr 87 | Ra 88 | Ac* 89 | Rf 104 | Db 105 | Sg 106 | Bh 107 | Hs 108 | Mt 109 | Ds 110 | Rg 111 | | | | | | | |

Key — H 1 — atomic symbol — atomic (proton) number

H -1

*The lanthanoids (atomic numbers 58–71) and the actinoids (atomic numbers 90–103) have been omitted.

# Further reading

Clark, Jim, *Calculations in AS/A-Level Chemistry* (Longman, 2000)

Cox, Michael, *Chemistry: A-Level and AS-Level* (Longman, 1990)

Hill, Graham & Holman, John, *Chemistry in Context* (Nelson Thornes, 2000)

Ramsden, Eileen, *A-Level Chemistry* (Stanley Thornes, 2000)

Roberts, Royston M., *Serendipity: Accidental Discoveries in Science* (John Wiley & Sons, 1989)

Strathern, Paul, *Mendeleyev's Dream: The Quest for the Elements* (Penguin Books, 2000)

120

Leabharlanna Poibli Chathair Bhaile Átha Cliath
Dublin City Public Libraries